The Southern Orchard
Month-by-Month

A monthly guide to fruit tree care

The Southern Orchard
Month-by-Month

A monthly guide to fruit tree care

Trey Watson

ISBN (paperback): 978-1-959823-00-1

To the customers of Legg Creek Farm,
thanks for supporting us!

Introduction

The southern United States has a long history of fruit and berry production. For thousands of years, wild grapes, persimmons, blueberries, blackberries, pawpaws, mayhaws, and black cherries yielded fruit for humans and wildlife.

Much of the landscape and the plants have changed or gone missing from the forests. But pockets of Eden-like forests and fields still exist here and there throughout the region. For those of us who live in the South, we have a unique climate for growing plants. Blessed with a long growing season, typically mild winters, and usually adequate rainfall (on average), the South is a perfect location for growing things. The usually sub-tropical climate allows for the cultivation of a wide assortment of plants.

This book isn't about the wild edibles, or the specifics of each cultivated fruit that grows here. This book is a guide to growing orchard fruits, month-by-month, in the southern United States. For our purposes, the southern United States includes the following states: Texas, Oklahoma, Arkansas, Louisiana, Mississippi, Alabama, Tennessee, Georgia, Florida, North Carolina, and South Carolina. It also includes parts of Kentucky, Missouri, and Virginia. This region of the United States generally has a humid subtropical climate. The area has enough similarities to make some generalizations and provide some specific guidance for growing trees in the region during each month.

It is my hope that this book will help you grow your own fruit. There are few things that taste better than tree-ripened fruit, and few things that give the sense of freedom and independence more than producing at least some of your own fruit.

How to use this book

This book has each month of the year listed, with detailed information on recommend activities in the orchard for that month. Refer to these chapters each month as the year progresses for guidance on fruit tree care during that time.

Following the guidance for December, there are additional chapters on fruit tree management, including one with some common questions we receive at Legg Creek Farm. These chapters provide a more in-depth look at particular orchard management questions and will hopefully be useful in any month.

There is also a section on homemade sprays and pest control methods that can be used any time of the year.

January in your orchard

If you're like us, you've had some pleasant weather and then some cold, windy weather, which is a sure sign that January is here. You don't even need to look at a calendar to know that we've started another year around the sun. Speaking of the sun – it's angled towards the south but it's at least staying up a few minutes more each day. It's a pretty good guess that the cold is going to continue for another couple of months.

Here are some things to do in your orchard this month:

Pruning

January is the time to prune in southern orchards. A lot of gardening books suggest spring, but for those of us who live and garden in the South, January is the best time to properly trim, train, and prune fruit trees. It's helpful to start off by pruning any suckers that have come up below the graft union.

Cut them to the tree trunk. Removing those suckers allows the tree to put its full effort into growing the upper, productive portion.

Prune peaches, cherries, apricots, and plums to an open center, like this diagram

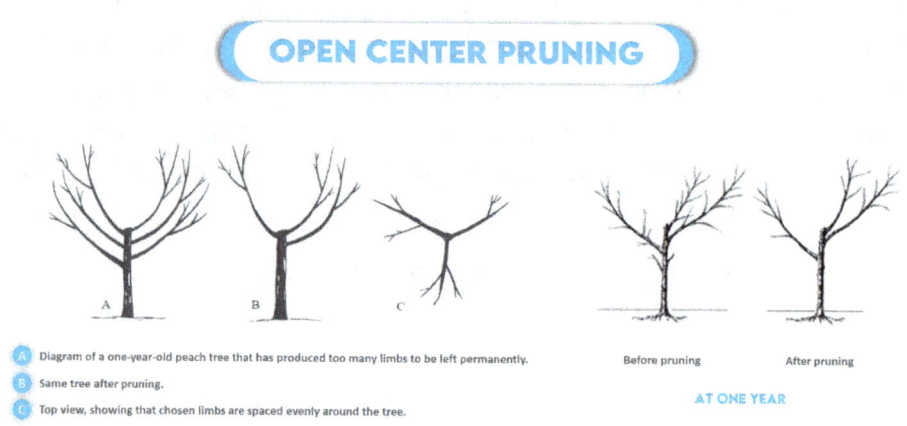

OPEN CENTER PRUNING

A   Diagram of a one-year-old peach tree that has produced too many limbs to be left permanently.
B   Same tree after pruning.
C   Top view, showing that chosen limbs are spaced evenly around the tree.

Before pruning   After pruning

AT ONE YEAR

Apples and pears can be pruned with a strong central leader, like this.

APPLE/PEAR TREE PRUNING

CORRECT

INCORRECT

Fig, persimmon, pomegranate, and native trees do not need pruning, except for removing dead or diseased wood.

Spraying

January is the optimal time for dormant oil spraying. Neem oil is a great organic, natural option, along with commercially available dormant oil sprays. Dormant oils, including neem, work by smothering overwintering insects. Coat every tree in the orchard with dormant oil or neem oil in January, and you'll have fewer pests bothering your trees during the growing season. Avoid spraying if it's predicted to rain in the next 24 hours.

Consider applying Bordeaux mixture to the dormant trees sometime in January. It's best to apply this fungicide once before the leaves bud out in Spring. Coat the bare trunks and branches with the mixture to prevent overwintering of fungal disease spores. Make sure the pruning wounds receive the mixture too. Copper-based fungicides are also effective and are usually sold in liquid form, making them easier to apply.

Mulching

If you haven't already done so, January is a good time to mulch the orchard. Hay, compost, pine straw (especially for blueberries), fallen non-fruit tree leaves, pre-made coconut coir round tree rings – each of these are a great mulch for fruit trees. Be sure to keep an inch or so gap between the mulch and the trunk of the fruit tree. A good mulch is 3-5 inches thick on top of the ground around the tree.

Planting/transplanting

Bareroot trees in particular do well when planted during this month. Fruit trees and other dormant plants can also be transplanted to other locations during this month. Planting during January takes advantage of warmer days that may start to show up; on any warm southern winter day, the roots of trees, including newly-planted trees, will grow and start to take root, even as the top stays dormant.

Make sure that the graft union is above the soil surface when the tree is planted.

February

February is usually a mixed bag in southern orchards, with frigid cold and then warming days all mixed together. The end of February typically signals the end of bareroot fruit tree planting, although the season for planting can be extended into early March, depending on the weather. It's a good idea to plant trees early in the month, which will allow the plants to take advantage of the warmer days before spring arrives.

Spraying

February is the month to continue with dormant oil/neem oil applications. Even if you've applied a horticultural oil in January, it is a good idea to apply it again in February. As the weather inevitably warms up on some days during this month, the insect pests will start to stir. An application of dormant oil, including neem oil, will help keep them in check when the weather warms up and the plants start growing. Do not spray horticultural oils on trees that are in bloom, as this will kill the blooms and subsequent baby fruit.

Final pruning

February is the final month for southern gardeners to prune fruit trees. Prune any dead or diseased branches, along with any branches or limbs that cross one another. Prune peach, nectarine, and cherry trees to an open center. Prune apples, pears, and Asian pears as mentioned and shown in January.

Fig trees and Japanese persimmon trees do not need pruning, unless desired by the landowner for aesthetic reasons.

Some trees will start budding and blooming this month, so it's usually best to prune before the leaves start sprouting.

Watch

Some lower chill varieties of apples, peaches, and pears will begin blooming this month, depending on weather conditions. Observe the blooms and watch for insect pests. Also watch for honeybees and native pollinators; an absence of pollinators will mean a lower fruit tree yield. On warmer February days pollinators should be stirring, which is a good thing for early apples and peaches. Avoid spraying any insecticides while the trees are blooming.

March

March is one of the most exciting times in the orchard! Many fruit trees are blooming during this month in the South. If late winter is mild, many varieties will at least start leafy growth during this month. Usually two to four weeks after bud break (when the leaves sprout), fruit trees will start blooming. Most fruit trees will have open blooms for about two to three weeks, during which time they will depend on pollinators to make fruit.

Spraying

Watch closely for pest infestations as the days warm and the fruit trees bloom. Avoid spraying anything on the trees while they're in bloom. On trees that have already bloomed, it's safe to apply an insecticide as needed for any unwanted insect visitors. Several days after the petals have fallen from the fruit tree blossoms, apply a copper fungicide to the trees, or use Captan. Particularly with peaches and other stone fruits, this initial application will help hold off the diseases that are so prevalent later in the season.

Another non-organic option is a systemic pesticide. This is a relatively new type of pesticide that is applied in liquid form, as a drench, around the base of the tree. The tree absorbs the pesticide, distributing it into its cells. The type of insecticide protects leaves, trunk, roots, and fruit from pests and many diseases, all with one application, usually in March. Systemic insecticides greatly reduce the need for weekly spraying for diseases on stone fruit and they are simple to apply. There are several approved brands in the United States, including Bayer, Bonide, and others. This insecticide has show promise, protecting fruit and trees in the climate of the south, with a single application each year. There are organic systemic pesticides that recently came on the market. These organic systemic pesticides use natural, plant-based acids and other compounds to protect the plant or tree, but they do not provide as broad-based protection as the conventional, non-organic systemic pesticide.

Weeding

March is a good time to start clearing away weeds that may have grown over late winter. Weed around young fruit trees to prevent weeds from competing with the fruit trees. Mature fruit trees also benefit from weeding around their base. Keeping tall weeds away from fruit trees prevents pests like grasshoppers from getting established in the orchard, where they will feast on succulent young fruit tree leaves.

Bagging fruit

If young fruit develop on trees during this month, it is a good idea to consider bagging the fruit to prevent pests from laying eggs or boring into the fruit. Mesh bags, paper bags, and even plastic zipper bags can work to reduce the need for pesticides on the tree by physically protecting the fruit from insects, some diseases, and animal predation. Late March is the time to at least analyze the need for fruit bagging.

Watch for pollinators

Pollinators of all types – honey bees, wasps, bumblebees, etc, - are essential for fruit production. In certain places and in certain years, the number of pollinators will be lower, for a variety of reasons. Many larger orchards allow beekeepers to move hives into the orchard over the spring. Other fruit tree growers have a few beehives for their own honey production and for pollination. If you see few or no pollinators in the orchard during spring bloom, consider offering your fruit tree space to a local beekeeper, or consider getting your own beehives to boost fruit pollination.

April

April is the month for fruit tree blooms and baby fruit! Trees that bloomed in February and March will be filled with baby fruit. Fruit trees that bloom this month will form fruit soon after the flowers fade. Pollinating insects should be active in the orchard on most days.

Spraying

Continue spraying as needed for insects and diseases. Fungal diseases may show themselves on warmer, humid days in April. Insect pests are also fully waking up on warmer April days. Spray trees as needed for insect pests. For insects, insecticidal soap is a good choice. It is always a good idea to avoid spraying any insecticide when the trees are blooming. Check the orchard for pests often. Apply a fungicide spray as often as once every other week for peaches and other stone fruit. Apples also benefit from fungicide spray at least once a month.

If a systemic pesticide was used, watch for any breakthrough infestation of insects or diseases, but otherwise there should be no need for additional spraying.

Bagging fruit

April is a good month to continue bagging fruit. Bagging fruit is a worthwhile way to reduce pesticide sprays while also growing nice fruit.

Fruit thinning

April is the best month for fruit thinning in the South. For many gardeners, thinning fruit is something they would rather avoid. The process of thinning fruit involves removing baby fruit. It's a painful process, but a necessary one. Many varieties of fruit trees naturally overproduce fruit. If left crowded on the tree, these fruit will all be small, lessening the edible yield of the tree while inviting disease. Fruit thinning is usually necessary with all fruit trees except native fruit trees. To thin fruit, remove anywhere from 1/3 to ½ of small fruit soon after they develop, carefully clipping them away at their stems. Chickens and other livestock appreciate the thinned fruit; with enough thinned small apples from the orchard, pectin for canning can be made at home.

Protecting fruit from wildlife

The young fruit growing on trees in April will start to invite birds and other wildlife to the orchard. The sugars and juices in the fruit will attract fowl and mammals alike. Bagging fruit helps to reduce issues from birds to some extent. Other animals, such as raccoons and possums, will be more bold in climbing trees and removing fruit, especially at night. Some orchard owners and gardeners have gone so far as to fence off the orchard with electric fencing. Reflective mylar tape, made especially for frightening birds, is somewhat effective in the orchard at reducing fruit lost from flying thieves. Metallic hardware cloth or mesh wire wrapped loosely around the fruit tree trunk will help keep some climbing pests away. A dog who runs free near the orchard goes a long way in keeping unwanted animals away from fruit trees.

May

May is the month when the orchard is in full production. During this month, grass is growing readily between trees and trees show have at least some fruit.

Fertilizing

With the warmer days of May, it's time to fertilize fruit trees. Most experts recommend fertilizing trees after they've grown a full year since planting. This is often a good idea, unless the young tree's leaves are turning yellow. Yellowing leaves, absent any other issues, is an indicator of a lack of soil nutrients. This is most often an issue with young trees planted in sandy soil with minimal organic matter or in heavy clay soil. It is safe to fertilize fruit trees that are showing signs of nutrient deficiency, even if they're newly planted.

It's best to fertilize based on a recent soil test. In the absence of that, use a balanced fertilizer such as 13-13-13. Alternatively, use a balanced organic fertilizer, such as composted animal manure or compost. For organic fertilizer, apply liberal amounts around the base of the tree, staying a few inches away from the trunk. Fertilize with commercial fertilizer based on the rates the manufacture recommends, which is usually based on the thickness of the fruit tree's trunk. May is the best month to fertilize for the first of two to three applications during the growing season.

Checking for pests and spraying

Pests and diseases will often start to thrive in May. Grasshoppers, aphids, and plenty of other pests will make their presence known in May if they haven't before. Check the orchard frequently, at least weekly and preferably daily, to look for signs of insect infestation.

Inspect tender young shoots and young fruit, looking for eaten leaves, damaged fruit, and other signs of pest damage. Spray as needed to control insects.

Diseases may also show up in May, thanks to warming days and higher humidity. For peach, nectarine, and cherry trees, continue to spray for diseases on a regular basis, unless a systemic pesticide was used earlier in the year.

Protection from predation

Mature fruit trees will have small fruit on them starting this month. Immature fruit is usually not appealing to mammals, but birds may take advantage of young fruit on trees. To prevent this, use bagging on young fruit at some point after thinning the fruit.

Reflective tape or other mammal or bird scaring devices can be hung on trees starting this month. None of these methods are perfect when it comes to saving fruit from predation, but they do reduce the amount of fruit lost to animals.

Mowing between trees/weed control

Grasses are usually growing rapidly during May, thanks to rainfall, warmer weather, and longer days. Keeping the spaces between fruit tree rows low will help reduce some insect pests and will discourage mammals from sneaking into the orchard.

June

Check for ripeness/harvest

Some fruits, including certain apple, peach, and nectarine varieties, ripen during this month. It helps to know the variety and the approximate time it ripens so that fruit on the trees can be checked for ripeness. Early peach varieties, often clingstone, will be ripe during this month. Check fruit for ripeness often – other creatures will also be waiting for the ripe fruit. Checking first thing in the morning will ensure ripe fruit isn't eaten by birds or other animals over the course of the day. Ripe fruit should come off the tree will a gentle tug but not a hard yank. Pull off one fruit to check for ripeness; if it's not ripe, try again in a few days. Dorsett and Anna apples will ripen during this month.

Check for pests

The warmth and humidity of the South is a breeding ground for pests. Common fruit tree pests to watch for this month include the following:

Grasshoppers – These big pests will eat leaves off of fruit trees. They don't generally eat enough leaves to bother mature trees, but they will completely defoliate younger fruit trees. They are especially attracted to fresh new growth. To protect fruit trees from grasshoppers, keep grass from growing up around young fruit trees. There are a number of effective insecticide baits for grasshoppers, but these often have little impact on populations. To protect young fruit trees, keep them well-fertilized and watered, weed-free, and apply Seven spray or dust, or other insecticide labeled for grasshoppers. Alternatively, blend a mixture of two cups of chopped garlic with 10 cups of water. Bring this mixture to a boil, and then let it sit overnight. Mix one part of this solution with three parts water and apply to plants. This natural mixture repels grasshoppers and other pests and works in the orchard. Malathion also works to control these pests.

Aphids – Aphids are attracted to younger growth on fruit trees in the spring and summer. These small insects will at times congregate on growing plant tips. Aphids are easily controlled by a spray of soapy water. There are numerous other insecticides that work for aphids. Aphids can cause deformity and death in new growth on fruit trees, but they are easily controlled.

Caterpillars – Caterpillars of all species will develop over the course of the spring and summer. The timing of caterpillar infections vary depending on the season and weather conditions. They will congregate on younger leaves, often on the underside. Careful observation of the orchard is essential to stop infestations.

Caterpillars can strip leaves from orchard trees quickly; repeated defoliation during the growing season can be fatal to fruit trees. Use *Bacillus thorengensus* (Bt) as an organic control method for caterpillars. Neem oil mixed with water is also effective, especially when applied in the evening.

Be on the look out for fruit pests during this month, and all summer. If the fruit is bagged, pest issues directly on fruit should be minimal. But if fruit pests do show themselves, an application of neem oil will help stop the issue while preserving the fruit for later harvesting.

Irrigate if needed

June sometimes begins the season of drought in the South. Irrigation is essential in the southern orchard, either from the end of a water hose or from an irrigation system with polyvinyl tubing. Water newly-planted fruit trees twice a week with a deep watering if there is no rain. Deeply water mature trees once a week during dry spells.

Rainfall is obviously ideal for fruit trees and all plants. When there is hot weather coupled with no rain, all fruit trees need to be watered. The bigger the tree, the more water they will need to produce sweet, juicy fruit. It's hard to judge exactly how much water is needed in a given situation. A general rule of thumb is at least an inch of water per week either from rain or irrigation.

Mulch

Check to verify that the mulch around the trees is still thick and preventing weed growth. If weeds are poking through, pull them out and add another inch or so of mulch.

July

In July, it's usually hot in the South. But things are growing like crazy, so there is plenty to do in the orchard.

Check for fruit ripeness

In July, continue to check for ripeness on plums, peaches, nectarines, and some varieties of apples, pears, and cherries. Figs may also ripen during this month. Many early apples are ripe in July, and many peach and plum varieties will be ripe this month.

Fruit protection

Keep pie plates, scarecrows, or holographic flagging tape on or around trees to protect the fruit from birds and other predators. As the fruit ripens, it will become even more attractive to other creatures who will want to get to the fruit first.

Spray

Continue spraying fruit trees during this month. Peaches and nectarines will need fungicide application until they ripen, usually applied every other week. Apples, plums, pears, and other fruit should be sprayed only as needed for pests and diseases.

Fertilize

July is a good month to fertilize the orchard. It's best to fertilize based on a soil test. July is the time to fertilize for summer fruit tree maintenance. Apply ("Side dress") a balanced fertilizer, such as 13-13-13, at the recommended rate on the fertilizer bag or based on the soil test. Temporarily remove the mulch around trees before applying a conventional fertilizer. An organic fertilizer can be added directly to the top of the mulch around the trees. Liquid fertilizer can be applied directly to the mulch without removing it. For any non-liquid fertilizer, water it in with irrigation, or apply it just before a rain.

Irrigate

If it wasn't necessary to water in June, most Julys in the South are dry and hot, so it likely will be now. Continue irrigating as needed during dry spells. Water trees deeply at least once a week in established orchards. Water any trees that show moisture stress and water younger trees more frequently, usually 2-3 times per week for best results.

Pests

Continue monitoring for pests and diseases and applying controls for them as needed. A well-watered orchard during hot, dry spells will attract insect pests.

August

August is the time of peak growth in many southern orchards. Healthy, happy trees that are thriving in our warm climate with plenty of water should be doing well this month.

Harvest what's ripe

Several varieties of fruit ripen during August, including some apples, pears, peaches, figs, and pomegranates. Check for ripeness frequently on these trees during this month.

Monitor for pests

Insects and diseases will continue to show themselves during August. Continue to monitor trees to insure no major pests or diseases are present.

Irrigate

The average apple is 84% water. August in the South is often on the drier side. Continue to water trees as needed if there is no rain.

September

September often heralds the beginning of cooler weather in the South. It's also a time of ripening fruit and shorter days. Hopefully, it's the month when the rain starts falling again if the summer has been especially dry.

Check fruit ripeness

Some varieties of apples, pears, cherries, and some native fruits and nuts are ripening during this month. Check fruit for ripeness and harvest ripe fruit before the birds and other wildlife get them. Pink Lady and Granny Smith apples will be ripe this month.

Plant cover crop

September is the month to begin planting a cover crop in the orchard. Native grasses can serve as a fine cover crop in many cases, meaning there is no need to sow additional seed for a cover crop. But in areas where there is bare soil, particularly on a slope, a cover crop can prevent erosion and improve soil health. A cover crop of legumes over winter can help build nitrogen in the soil at very little cost. Cover crops can also be planted as "trap crops" outside of the orchard. "Trap crops" bring animals, such a deer and other browsers, away from the orchard to prevent damage to trees over winter.

Planting a cover crop is optional in the orchard, but it does have benefits in many cases. Cover crops can be legumes, such as clovers, winter peas, or hairy vetch. Grasses, such as annual rye grass, can also be used. If livestock are involved in the orchard system, they can graze the cover crop at times over winter and into spring.

Avoid planting a cover crop near fruit trees that are younger than a year old; the cover crop can sap nutrients from younger fruit trees.

Clear leaves and fallen fruit

Fruit tree leaves and fruit may start to fall in late September. Rake and remove fruit and fallen leaves from the orchard to eliminate spaces for insects and fungal diseases to overwinter.

October

Days are getting shorter and the air smells like fall. The growing season is starting to wind down during the month of October.

Mulch for winter

October is a good month to add mulch around fruit trees. Mulching over winter prevents the rapid changes in soil temperatures that can accompany the alternating cold and milder temperatures we have in the South. It also helps keep moisture in the ground and creates a better micro-environment for beneficial soil microorganisms

Clear leaves and fallen fruit

Continue to remove fallen leaves and fruit from around the orchard trees. Burning fallen debris safely away from trees or removing them from around the fruit trees will help reduce diseases in the orchard by eliminating places where diseases and some pests overwinter.

November

The orchard in November is preparing for the long slumber of winter dormancy. The first freeze is typically during this month. Trees should be bare of fruit, with the exception of Japanese persimmons. Japanese persimmons will cling to branches after trees go to sleep, giving fresh fruit when other trees are dormant.

Continue to remove fallen leaves, and any fruit tree debris from around the trees. Keep the orchard watered if the weather is dry. If the weather gets cold enough to trigger dormancy during November, then this month is also a good time to plant fruit trees and berry plants.

December

It's the end of the year! The southern orchard should be fully dormant, though a few persimmons may still be hanging on trees.

Plant fruit trees

December is the beginning of prime berry plant, grapevine, and fruit tree planting time. Dig a hole big enough to accommodate the roots without crowding. Blend organic matter in at planting and mulch after the tree is in the ground, keeping mulch a couple of inches away from the new tree trunk.

Dormant oil

At least once during the month of December, consider applying a dormant oil, such as neem oil, to trees. Cover all the bare trunks with the oil. This process will smother overwintering pests and give your fruit trees a healthy start in spring.

Mulch

Make sure fruit trees have adequate mulch around them. Mulching helps reduce wide fluctuations in soil temperature and will protect the fruit tree roots if the temperatures get much colder than usual.

Fruit Trees Growing
Tips and Tricks

Chill hours

Chill hours are the number of hours it is below 45° F in a specific geographic area. USDA hardiness zones are helpful for selecting many plant varieties, but it is chill hours that help determine the best fruit trees for a location. Chill hours vary widely across states and can even be different in different locations in the same county. Chill hour averages, often measured by research universities, are available online for most of the U.S. It is best to select trees from a nursery that provides the number of chill hours for each fruit tree variety. Red Delicious, a popular apple that is commonly-planted and is widely available at supermarkets, needs around 1000 chill hours. Much of the South – from central Texas and east and south through Georgia, South Carolina, and Florida - experience fewer chill hours than that in an average year. This means that many popular and famous varieties, especially apples and pears, are not suitable for growing in many parts of the South.

Chill hours trigger a response in fruit trees and certain berry plants that causes the formation of healthy flowers. If a fruit tree that requires 1000 chill hours is grown where the average is only 600, the fruit tree will grow but it will not flower or produce fruit most years. The "average" year may seldom occur, meaning that some years may see 1000 chill hours and other years may see 300. But measured over time, that average number is a good enough guide for landowners to select fruit tree varieties.

Average chill hour maps are available for most states in the U.S. State-level maps allow the gardener to get more specific information on chill hours for their specific area. Please see Appendix A for county-level chill hour maps for the southern U.S.

There are two methods used to calculate chill hours. One is to measure all the hours below 45° F. The other method involves removing the hours below 32° F, since at that temperature the tree is simply protecting itself from the cold and not working towards dormancy. Depending on the year, this method reduces the total chill hours calculated by about 20%. For the home gardener, either method works and most research universities and state agriculture extension services use and publish data from both methods. The important thing is to select trees that have the proper chilling requirement for the area where the tree will be grown. Many online and mail order garden companies list USDA hardiness zones for fruit trees. USDA hardiness zones are the measurement of the coldest average temperatures for an area. This information is only marginally helpful when it comes to growing fruit trees. An apple tree that needs 1200 chill hours will be hardy in northern Louisiana, but it will usually not bear fruit, since there will rarely be 1200 chill hours in that region. For this reason, it is best to select fruit trees based on chill hours rather than hardiness zones.

Fruit tree fertilization

Fruit trees, like all growing plants, require nutrients for proper growth. Basic commercial fertilizer contains nitrogen (N), phosphorus (P), and potassium (K), usually marked on the package as percentages of those nutrients. N-P-K are essential for all plant growth and commercial fertilizer, such as 13-13-13, contains those nutrients in large amounts. In many cases, with actively growing fruit trees, commercial fertilizer helps give fruit trees what they need for sustained growth and fruit production.

But commercial fertilizer is not perfect. In addition to the expense of commercial fertilizer, it also usually lacks micronutrients. Micronutrients are present at some level in most soils. When absorbed by plants, they help enhance certain essential functions. In fruit trees, the micronutrients impact flower and fruit development. Fertilizers that are derived from natural sources, such as compost, animal manure, and others, usually have adequate micronutrients for fruit trees. A combination of commercial fertilizer, with organic or natural fertilizer as a side dressing, does a good job of fulfilling fruit tree nutrient needs. Of course, plenty of entirely organic options are also available, allowing the orchard owner to grow a completely organic orchard.

A compost blended of kitchen scraps and garden clippings is a perfect source of micronutrients in the orchard. Apply compost beneath a layer of mulch in spring and summer around each fruit tree.

Animal manure should be composted before use in the orchard. Composted oultry litter is high enough in the three main plant nutrients that it makes a great all around fertilizer in the orchard.

Orchard Soil

Soil testing, conducted by a private or university laboratory, is a helpful tool for knowing just how much fertilizer to apply. Much of the southeastern United States has soils that are naturally acidic, meaning that most soils in this region lower on the pH scale. A proper soil pH level is essential for making all plant nutrients available to the fruit trees. Even if a soil has the right nutrients, a pH that is too low or too high will cause certain nutrients to not be absorbed by a fruit tree's roots. A pH level of 6-7 is optimal for most fruit trees. That's also the level where most nutrients are available for fruit trees and other plants. Fruit trees can tolerate a wider variety of pH levels, but the range of 6-7 is ideal. Agricultural lime can be applied to raise the soil pH to the correct level for the fruit orchard. Wood ash is another source of higher pH material that will raise soil pH while also providing potassium and micronutrients to fruit trees. In acidic soil, blending a healthy amount of wood ash with compost and mixing it with native soil at planting will often help raise soil pH close to the optimal level.

A well-drained, sandy or sandy loam soil is ideal for most fruit trees. With some effort, specifically with adding organic matter and ensuring proper drainage, fruit trees can be grown in clay soil. The number one requirement for fruit tree soil is proper drainage. Well-drained soil of any kind can be made to grow fruit trees.

Thinning Fruit

Most modern fruit trees are "excessively" productive, meaning they will produce more fruit than the tree can reasonably sustain. Some fruit trees, such as apples and pears, tend to produce fruit at the tips of branches. Peaches, nectarines, and other stone fruits often grow along the branches at various points. In either case, it's necessary to remove around half of the immature fruit in order to get large a harvest of high quality fruit. Thinning fruit helps the remaining fruit grow to a full size. It also prevents tree branch breakage due to excessive weight and it helps prevent disease by allowing air flow among the fruit. It can be a painful process to remove so much baby fruit, but it is necessary. Immature apples that are harvested by thinning can be made into pectin for home canning. Other thinned fruit can be fed to livestock or added to the compost pile.

Animal predation

Fruit growing on trees, especially as it nears maturity, is a tempting target for fur and feather-covered creatures. Pests as varied as mice, deer, raccoons, and all types of birds are drawn to fruit on trees. For those of us fortunate enough to live near the woods, this problem is persistent. Thankfully, some control measures can reduce loss of fruit to animals.

Mice/Rabbits — Field mice will nibble on young fruit tree trunks, sometimes girdling and killing the tree. To prevent this, use kitchen foil or a commercial tree wrap around young tree trunks. Only the bottom 1-2 feet of the tree, including the graft union, should be covered in foil. The foil discourages mice and rabbits from gnawing on the tree trunk. It's also helpful to keep weeds from growing up around the trunks. This removes hiding places for rodents near the tree trunk.

Raccoons/possums — These furry creatures will snag fruit straight from the trees. Some of this is unavoidable. Having a dog to roam the orchard helps, as does keeping brush cut down around areas where fruit trees are planted. Hanging aluminum pie plates from tree branches will help to some degree, as the wind causes the pie plates to move around and make sounds.

Deer — Deer are often interested in the young shoots and leaves on fruit trees. In the early spring, young bucks may rub their growing antlers on fruit tree bark, often harming or killing the tree. They will also nibble on young fruit, consuming or destroying them. Many fruit trees growers fence off their orchard, hoping to keep the deer out. Some gardeners even put fences up around each individual tree.

Fencing certainly helps with keeping out deer, but it is also expensive. Commercially available deer repellents are often effective, especially if applied to fruit trees every other day for the first two weeks in early spring and then applied once a week after that during the growing season. An effective, homemade animal repellant spray is listed in the chapter on spraying.

Some gardeners have found success with solar-powered electronic devices that emit a high-frequency sound that deter deer. The same devices can also deter other mammals. The efficacy of such devices seems to vary. If they work, they usually repel most mammals that do harm to fruit trees.

Birds — Birds are every commercial orchard's worst nemesis. There is no way to have 100% protection for fruit trees when it comes to birds. In commercial orchards, "boom cannons", devices that create a loud, explosive sound at either random times or when birds are near, are used to scare away the fruit feasting flyers. These sounds devices work to some extent, though they have to be moved and adjusted to keep the birds from getting used to them.

For home orchardists and gardeners, people have found varying degrees of success with holographic reflective tape, metallic pie plates hung by strings, fake owls, and scarecrows. The holographic reflective tape is also used in commercial orchards. It does seem to help protect much of the fruit on trees. Bagging individual fruit also helps protect it from birds and other pests.

Irrigation methods

If you grow fruit trees in the South, you will need to irrigate your trees. It is guaranteed that we will have dry spells, including the possibility of multi-week or months-long droughts in the middle of the growing season. In the home orchard, irrigation is usually accomplished with a water hose. A drip irrigation system, usually attached to a regular faucet, is a cost-effective and water-conserving method of making sure fruit trees get the water they need when the weather is dry.

Polyvinyl tubing of various diameters can be purchased and laid out in the orchard, with drip locations (emitters) placed at the base of each fruit tree. The one disadvantage of permanent drip irrigation is the tubing is laid on the ground between trees, making it difficult to mow between trees due to the risk of mowing over the tubes and destroying them.

Go-to Sprays and Pest Control Items for the Orchard

It is a good idea to have the following items on hand when growing fruit trees in the South:

Boudreaux mixture — This is the most effective organic fungicide for the home orchard. It's an old fungicide, with more than 200 years of history behind it. It is a combination of copper sulfate and lime, and is usually sold as a powder. In the orchard, it should be used as a preventative rather than a cure, or to prevent the spread of the already existing disease. It does work for a number of apple, peach, and plum diseases, including issues that plague peaches in the South.

Copper fungicide — An old standard, copper fungicide is an effective organic fungicide for a wide variety of fungal diseases on peaches, apples, pears, and other fruit trees. It's a "first line," go-to fungicide that is broad spectrum enough to help with most fungal diseases. It is also effective on grape diseases. Copper fungicide is usually sold at large garden centers already mixed in a spray bottle. It is also sold as a concentrate that can be mixed and applied with a spray tank.

Neem oil — A type of dormant oil, neem oil is not derived from petroleum like traditional dormant oil, but still functions the same way. It can be purchased as a concentrate or a ready-mixed spray. Neem is a great natural spray. Spray in early morning or late in the evening for best results. Spraying in the middle of the day may burn fruit tree leaves. Neem oil can be used throughout the growing season. It is a broad-based insecticide and a fungicide for several fungal diseases. It is also safe for animals and the environment. Avoid spraying neem oil on trees while they're in bloom.

Spinosad — Many variations of Spinosad are approved for use in organic agriculture. Spinosad is created from specific soil bacteria and is effective against a large number of insect pests, including caterpillars. It's an effective organic solution in orchards.

Malathion — If an organic solution doesn't work in dealing with insect pest infestations, malathion will do the trick. It is relatively short-lived in the orchard, and needs to be reapplied every seven days. Malathion is a go-to insecticide in commercial orchards. Malathion is harmful to bees and other beneficial insects, so this pesticide should be used sparingly.

Captan — This chemical fungicide works wonders at controlling fungal diseases on peaches, plums, and apples. It is effective and can be kept on hand in case fungal diseases become uncontrollable. Peaches will often get brown rot and the fruit will be inedible; Captan is the fungicide that treats this disease.

Wettable Sulfur — Available at many feed stores and garden supply stores, this organic fungicide is effective on a wide range or fruit trees and other plants. Wettable sulfur is still effective, even after thousands of years of use in agricultural settings.

Balanced fertilizer — A commercial fertilizer such as 13-13-13, is convenient to have on hand for fertilizing the orchard a few times a year. An organic substitute, including animal manure and compost, is also effective.

Simple homemade pest remedies

Some orchard and garden sprays can be made at home, with materials that can be easily grown or purchased.

General Insecticide

If insect pests are infesting the orchard, this simple, general purpose insecticide will do the trick. It is effective in killing all insects that receive the spray directly, including beneficial insects. Applied alternatively with the garlic spray, listed below, this insecticide will control grasshopper infections in the orchard.

Ingredients:

>   1 and ½ tablespoons dish soap
>
>   1 table spoon cooking oil (vegetable, corn, olive, etc.)
>
>   1 quart of water

Directions:

Mix all ingredients together in a jug, spray tank, or spray bottle. Apply as needed to control insects in the orchard.

Garlic spray/repellent

This is an odiferous and effective spray for controlling some insect pests and repelling others. It is best used no more than once a week during the growing season. Avoid spraying while the trees are in bloom.

Ingredients:

5-6 bulbs of garlic (homegrown or store-bought)

1 gallon of water

Cheese cloth or clean used t-shirt

Directions:

Break bulbs of garlic into cloves and peel the dry skin from the cloves. Place half the garlic in a food processor or blender with 1 cup of water. Chop/blend well, then pour that concoction in a separate bucket or bowl. Blend or chop the other garlic cloves in the same way. Mix all ingredients in a large bucket. Let the mixture sit covered overnight, then strain the mixture into another bucket through cheesecloth or a clean t-shirt. The mixture is now ready to use and can be placed into a spray bottle or spray tank for application.

Pepper spray

Hot pepper spray is an effective insecticide on smaller insects, and a deterrent to others. It is also an effective protectant from mammals, since the spicy flavor of this spray is not liked by rabbits, squirrels, deer, and other animals. This is an intense spray, and should be prepared with gloved hands. Avoid contact with skin and do not rub eyes while preparing this spray. Apply the spray in the evening, to avoid burning tree leaves during the heat of the day.

Ingredients:

   2 cups (chopped) hot peppers (Jalapeno, Habanero, Cayenne, etc.)

   1 gallon of water (less for a more intense spray)

   2 tablespoons of dish soap (optional, but helpful to get immediate kills on some insects)

Directions:

Carefully, with gloved hands, chop the peppers into chunks with a knife. Add peppers to a sauce pan with a little water (removed from the 1 gallon). Simmer peppers for 10-15 minutes, then add remaining water and bring to a boil while covered. Take off heat and leave covered, ideally leaving to cool overnight. Strain out the peppers and mix in soap. Pour into spray bottles.

Powdery Mildew Spray

Powdery mildew can be a problem on certain varieties of orchard trees. This fungal disease is especially common in the warm, humid days of a southern summer. It may also show up in the early fall, particularly if the summer has been dry and the rain has returned. Powdery mildew can infect many types of plants, including blueberries and garden vegetables. There are chemical remedies for powdery mildew, but these are often unnecessary. Try this easy, homemade remedy instead:

Ingredients:

1 gallon of water

1-2 tablespoons of baking soda

½ teaspoon of dishwashing soap

Directions:

Mix all ingredients together in a gallon jug or other container. Pour into a spray bottle or other spray container and apply to top and bottom of infected leaves. Reapply after rain.

Common Fruit Tree Cultivation Questions

How do I plant fruit trees?

Fruit trees usually come ready for planting in two ways: containerized or bareroot. In the past, ball and burlap was a common method for transporting trees to their final planting location, but that has fallen out of favor.

Containerized fruit trees can be planted virtually any time of the year, though trees planted in the heat of summer need extra mulch and watering to ensure their survival. Container trees are usually more expensive and can sometimes be "pot-bound" with roots that are so crowded in the container that they take longer to get established in soil. The advantage of container trees is their wider planting season window. To plant container trees, dig a hole just a little bigger than the container and gently remove the tree and the soil. This is easy to accomplish by laying the tree over on its side and pulling it out by the stem. Water the tree thoroughly after planting.

Bareroot trees have the advantage of being less expensive than container trees. They have a shorter planting window, usually November-March in the South. Bareroot trees are often available from mail order and online nurseries. Many additional varieties are usually available as bareroot trees versus container trees. To plant bareroot trees, dig a hole big enough that the roots can be spread out without crowding. Place soil around the roots and gently pack it in. Water thoroughly, and then add more soil to any places where the soil has settled in the hole. Plant with the graft union a couple inches above the soil line. ***Winter weather and severe cold does not hurt most newly planted bareroot fruit trees as long as the roots are***

completely covered by soil. Figs and pomegranates are the exception; the tops of those fruit trees will die back in extreme winter cold.

What are chill hours?

Chill hours are the number of hours below 45° Fahrenheit. For university research, a temperature range of 32-45° is used to measure chill hours. But by and large in the nursery industry and with home gardeners, any temperature below 45°F when the trees are dormant is counted towards chill hours.

Chill hours over winter are necessary for fruit trees to produce blooms the following spring. Cold weather for a certain number of hours triggers what's called *vernalization*. Vernalization refers to the tree's process of generating flowers the in spring. With inadequate cold, as measured in chill hours, fruit trees cannot create blooms. Without fruit trees blooms, there will be no fruit.

How do I determine chill hours for my area?

To determine the chill hours for your specific locations, please refer to the county level maps included in this book. For further information, contact your local county extension agent.

What is the best soil for fruit trees?

Any well-drained soil is suitable for fruit trees. Clay, sand, loam, or any combination of the three main types of soil will grow fruit trees, provided that the soil drains water after rain or irrigation. Clay soil or soil that is predominately clay is usually the least well-drained soil type. As long as the water does not stand on the soil surface more than a few hours after a heavy rain, the soil should be well-drained enough for fruit trees. Flat or level clay soil is usually the worst for planting fruit trees; if you have clay soil, it is better to plant the fruit trees on a sloped part of the land.

Sand and loam soils drain water better than clay, even on a flat piece of land. This is the advantage of these soil types; the only disadvantage is that sandy soil usually needs more frequent watering and fertilizing due to rapid drainage and leaching of nutrients.

Overall, all three soil types will grow trees, but more issues will arise from clay soil than from others.

What is the optimal soil pH for fruit trees?

The optimal soil pH is 6-6.5. pH is the measure of acidity in a substance. Soil pH is most accurately measured in a university or private soil testing lab, though a number of soil testing probes and kits are available for home gardeners online and in stores. Most of the soil in the South tends to be on the acidic side, though "black land" soils in Texas, Oklahoma, and a few other places are naturally less acidic.

Can I change the pH of my soil?

If you find that your soil pH is not optimal for fruit trees, there are a couple of things you can do. If your soil is too acidic, as shown by a low pH, adding lime to the soil can help raise pH to an optimal level. Lime can be broadcast over the entire row where fruit trees are to be planted, or it can be mixed into the soil at each planting site. There are a few different types of lime, but each can be applied to fruit tree planting areas. Apply lime based on soil test recommendations or the recommended rates on the lime packaging.

Soil pH can be lowered on sites where it is above 7 by using a fertilizer with sulfur in it. Vinegar can be used to lower soil pH in a specific location by mixing one cup of vinegar into a gallon of water. Apply this over several square feet and give the soil a shallow till.

Which trees do I need a pollinator for?

Peach and nectarine trees are all self-pollinating and do not need another tree for pollination.

Apple trees usually need to be pollinated by another apple variety that blooms at the same time. A handful of apple varieties – Anna, Dorsett, and a few others – are self-pollinating. Other apple varieties are partially self-pollinating but fruit yield is increased when planted with another apple variety. Granny Smith and Jonathan are a couple of apple varieties that are partially self-pollinating. Otherwise, consider planting at least two types of apple trees that bloom near the same time to ensure pollination.

Plum tree varieties can be either self-pollinating or not. Methley, one the best southern plum trees, is self-pollinating and it also pollinates other Japanese plum varieties. Many other plum varieties do need a pollinator.

Pear trees generally need another pear tree nearby for pollination. Anjou, Bartlett, and Keiffer are all self-pollinating pear varieties. Other pear varieties need another pear tree that blooms around the same time for pollination.

Most Japanese persimmon trees are self-pollinating.

Most cherry trees require another cherry variety for pollination.

Fig trees are all self-pollinating.

The seller of the fruit trees should specify which fruit tree varieties need pollinators, and which varieties to use as pollinators. This information is usually listed on picture tags, catalog pages, or websites.

How do I know which trees work best in my area?

Selecting trees that have appropriate chill hours for your area is a good place to start. Local county extension agents can help with this, as can the internet and local farmers/gardeners. Prioritize selecting trees that need the correct number of chill hours for your area, and then verify they will do well in your USDA hardiness zone.

When is the best time to plant fruit trees?

The best time to plant bareroot fruit trees is when they are dormant, which is normally November through March. Containerized fruit trees can be planted most of the year, though it is often a good idea to avoid planting them in the heat of summer. The trunk of most freshly-planted bareroot fruit trees should suffer no issues in cold weather over winter.

Do I need to spray my trees?

By and large, fruit trees will need spraying for insects and diseases. For certain types of trees and certain seasons, this will be only as needed. Apple and pear trees in the South both seem to need less spraying than stone fruits. Pomegranate, fig, and persimmon trees only need to be sprayed when they are suffering from severe insect or disease infestations.

There are spray schedules in chart form in this book. They should provide guidance on when to spray and what to spray on a particular type of fruit tree. Your local county extension agent can also provide guidance on a spray schedule for your specific area.

How do I fertilize fruit trees, and when?

Established fruit trees are usually fertilized 2-3 times a year, usually once in early spring and once in mid-summer. Please see the month-by-month guide included in this book for more details on fertilizing.

How do I stop deer from eating my fruit trees?

Short of installing expensive electric fencing, spraying either a purchased or homemade deer repellent is usually the only way to keep deer from munching on fruit trees. Larger, more established trees are less often seriously harmed by deer. It is the younger, newly-planted trees that usually suffer. Deer will sometimes munch on the top of the new growth on younger trees; if this happens more than a few times, the young tree will be permanently stunted or killed.

Is it better to plant fruit trees on a slope?

This is a hard question. On one hand, you do want fruit tree soil to be well-drained, and sloped land is usually well-drained. As long as the soil isn't eroding, sloped land is a fine location for fruit tree planting, though it is not essential to plant on a slope.

Which fruit trees are easiest to grow organically?

In the South, pears, apples, figs, pomegranates, and Japanese persimmons are the easiest trees to grow organically. Well-selected apple trees for a specific area can be grown largely with organic practices. Pears, figs, pomegranates, and persimmons tend to need minimal spraying except for insects on occasion.

Peaches, nectarines, and cherries are harder to grow organically, but it is not impossible. Redhaven and Elberta are the easiest peach trees to grow organically. Stella cherry is the easiest cherry to grow organically in the South.

What is this brown stuff on my peaches?

Peach trees are notoriously disease prone, so it is probably brown rot. Consistent spraying of peach trees will help prevent diseases such as brown rot.

Why is my fruit small?

All tree fruits are more than 50% water. Small fruit is often the sign of a lack of water. It could also be an indication of low soil fertility. In either of these cases, the fruit tree will try to save itself by reducing the size of the fruit. To prevent small fruit, make sure fruit trees have adequate water and fertilization over the growing season. Also make sure the fruit tree has been adequately thinned; too many fruit on a tree will make all the fruit smaller.

How should I prepare the soil for fruit trees?

Soil is important for fruit trees, especially since they will be growing in that one place for many years. Mixing in organic matter with native soil is a good idea, at no more than a 50/50 ratio of organic matter to native soil. Additional compost or other mulch can be added to the top of the soil and allowed to rot, improving the soil over the years the fruit tree is productive.

When should I prune my fruit trees?

Please see the pruning chapter for this information.

Can I grow fruit trees from seed?

It is a long process, but it is possible to grow fruit trees from seeds. Thanks to centuries of breeding, the resulting tree will not be of the same variety as the fruit it came from. But there is a chance that the fruit grown from such a plant will be different and of high quality, making it worthwhile for those who have the time. It does take a long time to get a seed-grown tree to fruit production age – usually around 10 years. Most temperate (non-citrus) fruit tree seeds need 60-90 days of cold treatment to trigger germination. This can be done in a refrigerator and then the seeds can be planted outdoors in the ground or in pots.

How far apart should I plant my fruit trees?

Plant standard fruit trees 15-20 feet apart, with rows at the same spacing. Dwarf and semi-dwarf fruit trees can be planted closer, at 5-6 feet and 10 feet respectively. These spacings are by no means absolute; dwarf and semi-dwarf trees can be planted a bit closer if space is an issue, or farther apart if desired. Standard peach and plum trees can be planted closer than the 15-20 foot spacing and the trees can be pruned to keep them smaller to prevent branches of different trees from growing into each other.

How long do fruit trees live?

A properly cared for fruit tree will grow and produce for years. Pear, apple, persimmon, and fig trees will produce fruit for decades, and maybe longer. There is a pear tree still growing in Massachusetts that was planted in the 1600's. Peach, plum, and to a certain extent cherry trees are usually shorter-lived. There are exceptions with cherries and plums, with certain varieties producing for decades. Peach trees will live for 15-20 years or more, but their fruit yield generally declines after 10-15 years. Commercial peach orchards often replant trees after 10 or so years. Persimmon trees can grow and produce for one hundred years or more.

How should I care for newly planted fruit trees?

During dry weather, water newly planted fruit trees about once a week before they break dormancy. Make sure the soil drains completely within 24 hours, or the roots may rot. Keep the newly planted tree mulched. Avoid mulching young trees with sawdust, as this will absorb available nitrogen and keep it from the fruit tree. Keep weeds cleared out from around the tree. Water the fruit tree more frequently during dry weather and over the summer. Fertilize newly planted trees only if they show yellowing leaves or poor growth. A liquid fertilizer can be a quick boost for a tree planted in poor soil, as can any number or organic fertilizer applications. Don't be too aggressive with the fertilizer on new trees. Start fertilizing on a regular basis in the second year. During extended dry spells (often in July and August in the South), water new trees deeply at least twice a week. More mature trees will also need water, but young trees are particularly susceptible to drought stress. A tree planted in January will struggle to survive our harsh Southern summer dry spells without plenty of water.

Are there any companion plants for fruit trees?

Wildflowers and almost any plants that attract pollinators are excellent companion plants for fruit trees. Between rows of trees, wildflowers attract pollinators that will help increase fruit yields in the orchard.

There are herbs that some people use to repel insects from fruit trees, but on a larger scale it is hard to include herbs in a fruit tree planting. Herbs such as chives, society garlic, and rosemary are sometimes used to discouraged insect pests. In a planting of just a handful of trees, herbs can be planted between trees or between rows in beds. Some gardeners have reported that marigold flowers help reduce pests with fruit trees; these can also be added around smaller plantings of tree-bearing fruit.

Can I grow fruit trees in containers?

Most fruit trees can be grown in containers, at least for a little while. Standard (non-dwarf) fruit trees will outgrow their containers in just a couple of years, meaning they will become pot bound and need larger containers frequently. Fruit trees grafted on dwarf rootstock, however, can be grown longer-term in containers, as long as the gardener is willing to upgrade the container size every few years.

Container fruit trees need adequate water, fertilization, and chill hours, just like their field-grown kin.

While outside of the purview of this book, many citrus tree varieties make fine container trees in the South. They can be moved into a warmer, sheltered place over winter, allowing the gardener to grow them for years in areas with colder winter temperatures.

What rootstock is best?

Rootstock refers to the variety (usually non-fruiting) of tree that the better fruiting variety, called the scion, is grafted onto. There are numerous rootstocks that impart certain characteristics to the scion, including increased adaptability to certain soil types and resistant to some pests and diseases. The best rootstock for a given area depends on many factors. A local county extension agent can help with this. It is often better to ensure that the scion works for your area, especially in terms of chill hours and pollination, than to worry too much about the rootstock.

What is my soil type and is it good for fruit trees?

There are three major soil texture types: sand, silt, and clay. All soils are made up of these materials in various proportions. Soils with more sand tend to be better-drained, but they dry out quickly and often leach nutrients below the rootzone of plants. Silt is the leftover deposits of waterflow. It tends to grow plants well, but can be slick when wet. Clay soil has tiny particles which pack together closely, especially when wet. It holds water, which can be an issue with fruit trees. But if the soil drains, like on a hill side or any sloped land, then fruit trees can grow in clay soil. Clay soil also holds water longer as the weather turns dry, keeping moisture in a few days longer than other soil types. Loam refers to soils with an optimal mix of these textures. Loam soil is ideal for growing any plants. If you have loam soil, then you are in luck! There is less work needed to prepare loam soils for fruit trees. But soil of all textures can be made to grow fruit trees.

Orchard Floor Management

Orchard floor management is the process of managing the ground beneath and between your orchard trees. Properly managed, the orchard floor can provide a safe and nourishing habitat for your fruit trees. Orchard floor management includes:

- Mulching with organic materials
  (not black plastic or landscape fabric)

- Weed control

- Alley management (space between fruit tree rows)

- Cover crops

- Tree root microclimate management

Holistically, these activities create orchard floor management. Many of these concepts have been explored elsewhere in this book, but they all come together in orchard floor management.

To properly manage your southern orchard floor, start by mulching trees with an organic material. In the climate of the American South, black plastic or landscape fabric used as a mulch around trees is going to heat the soil around the roots too much, potentially causing tree death. Basic organic materials, including compost or woods chips, are effective mulches when applied to a depth of 3-4 inches above the soil surface around trees. Wood chips usually last 2-3 years and another, thinner layer can be added to them each year to maintain the mulch. Mulch keeps soil temperatures considerably lower in the summer, benefiting trees in our hot, humid climate.

Weed control around fruit trees of all ages is essential. For young trees, weeds growing directly at the base of the tree absorb vital nutrients from the young trees. In both young and older trees, weeds also provide a habitat for rodents, who enjoy gnawing on fruit trees of all ages. Brutally remove weeds from around trees, including between trees in each row. Weeds can be pulled, mowed, or sprayed with herbicide to remove them.

Alleys between rows of trees should be maintained in some type of permanent vegetation. Keep this area mowed during the growing season to keep away deer and rodents.

Mint, rosemary, garlic, and onions have been shown to repel rodents. These plants are probably safe to plant near fruit trees as companion plants.

Finally, when pruning fruit trees, remove the pruned material and any inedible fallen fruit from the orchard site and either burn it or bag it up and through it away (or feed the fruit to your chickens!). This will prevent some diseases and pests by removing their overwintering locations.

Fruit tree spray schedule for the South

There are a number of spray schedules online for commercial orchards in the prime orchard fruit producing areas of the United States. For home orchards in the South, try these spray schedules, divided into fruit types and organic and conventional.

Peaches, plums, apricots, nectarines – organic

Dormant - Neem oil or other horticultural oil

Budbreak (when tree leaf buds start to turn green) – Sulfur-based organic fungicide and neem oil

Petal fall – Neem oil or homemade pest spray; sulfur-based organic fungicide

Use sulfur-based spray as often as is allowed based on product label. until harvest. Spray for insect pests only as needed.

Apples, pears (including Asian pears) - organic

Dormant – Neem oil or other horticultural oil

Budbreak - Sulfur fungicide or Bordeaux mixture (for varieties not resistant to fireblight)

Petal fall – neem oil

Only treat orchard trees if needed for diseases and insect infestations. Wettable sulfur, Bordeaux mixture, and neem oil should be all that is needed to control pests and diseases in apple and pear trees. The homemade insecticides in this book can also be used to control insect pests.

On figs, pomegranates, Japanese persimmons, and native trees, treat for pests and diseases only as needed.

Common pests and diseases in the orchard include:

Apple scab – This fungal disease can hit apples at any point in the growing season. It can be a problem in southern orchards. The disease is characterized by a dry, scab-like texture on leaves and eventually black, deforming spots on apples. It will also defoliate trees, reducing or eliminating fruit from the tree for that season. Repeated defoliation will kill a fruit tree. Treat this disease with Bordeaux mixture. The initial application of Bordeaux mixture or wettable sulfur (fungicide) at budbreak helps prevent this disease. Apple scab can also infect pear trees.

Powdery mildew – This typically non-fatal disease will reduce fruit tree yield and weaken the tree, making it more susceptible to other diseases and to drought and heat stress. It can be treated with the homemade mix in this book, or with sulfur fungicide.

Fire blight – This bacterial disease of apple and pear trees can be controlled early in the season with a spray or application of sulfur fungicide. The disease reduces fruit tree yield and weakens trees.

Brown rot – This disease of peaches and nectarines causes the fruit to form brown mold and then rot; it is controlled by regular fungicidal sprays, applied before the disease appears

Root rot – Caused by a variety of fungal diseases natural in soil, fruit tree root rot occurs in trees that are planted in soil that isn't adequately drained. Plant trees in well-drained soil to avoid this disease.

Black rot – This disease of apples and pears is characterized by bruising, rotting lower ends of fruit. Good orchard floor management helps reduce black rot in fruit. A good general purpose fungicide, applied early in the season, will reduce the incidence of black rot.

Cedar apple rust – This disease causes defoliated apple trees, weakening the tree. Native cedar trees carry the disease. Fungicide spray can help reduce the disease, but the primary way it is controlled is by removing cedar trees from any location near the orchard.

Gummosis – Gummosis is the gum-like jelly that appears in the spring and summer on peach trees and other stone fruit. It is caused by a fungal disease that damages woody tissue, causing sap from the tree to leak. To control this disease, remove any dead wood from the tree, and protect the trunk from injury. Clear ooze means the disease is minimal; dark ooze means the disease may be more systemic. Spray the entire tree with fungicide early in the season to reduce the incidence of this disease.

Appendix A

These county level maps are based on the best available data on chill hours. They may not be accurate year-to-year, but they should give a decent estimate of chill hours in each county shown.

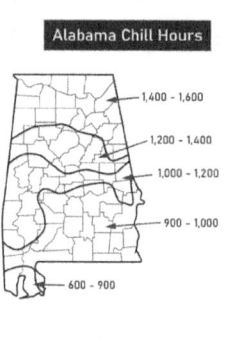

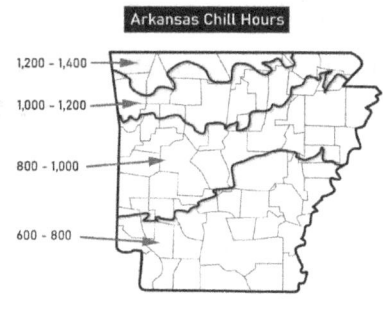

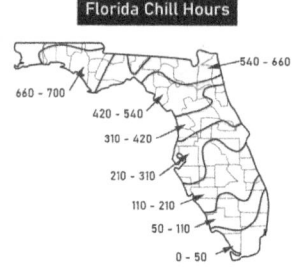

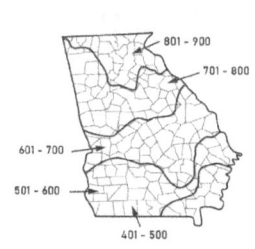

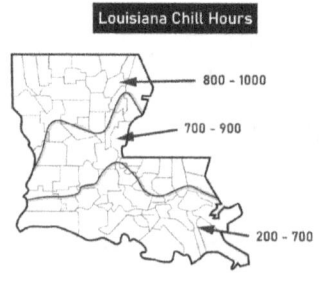

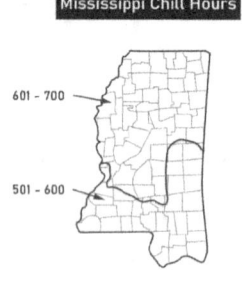

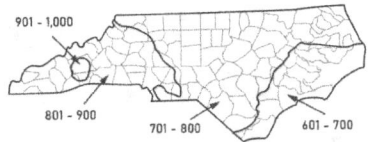

North Caroline Chill Hours

901 – 1,000
801 – 900
701 – 800
601 – 700

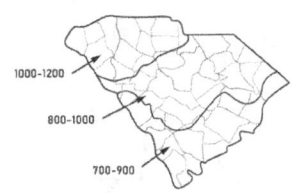

South Caroline Chill Hours

1000-1200
800-1000
700-900

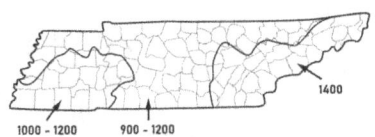

Tennessee Chill Hours

1000 – 1200
900 – 1200
1400

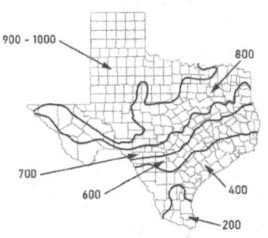

Texas Chill Hours

900 – 1000
800
700
600
400
200

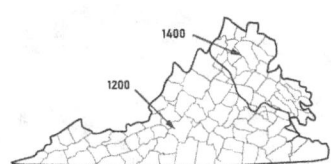

Virginia Chill Hours

1400
1200

Acknowledgments

Gardening has been a lifelong passion of mine. Thanks to LeggCreekFarm.com, I've been been able to share that passion with over 10,000 fruit tree customers. To each of those customers, I want to say thank you. Thanks for staying with us as we grew and generally being the most fantastic customers anyone could ask for.

I'd like to thank my wife for editing this book. She has an excellent perspective and her attention to detail is second to none. Plus she's beautiful and funny.

Thanks to my parents for allowing me to follow my passion starting at a young age. They encouraged me to enjoy gardening and fruit trees even as a child.

And God said, "Let the earth sprout vegetation, plants yielding seed, and fruit trees bearing fruit in which is their seed, each according to its kind, on the earth." And it was so. - Genesis 1:11

Email me!

Thanks for reading! If you have any questions or comments on this book, please feel free to reach out to me at leggcreekfarm@gmail.com. You can also find us on Facebook and Instagram at @LeggCreekFarm. You can find our website, where we sell lots of southern-adapted fruit trees, berry plants, grapevines, and native plants, at www.LeggCreekFarm.com.